How to Improve Your Corporate Identity

The Businessman's Guide to Creating a Better Company Image

Also by the Author:

Corporate Identity Manuals

Letterheads (Vols. 1-5)

Designing Corporate Identity Programs for Small Corporations

Evolution of Design

Logo International

Designing Corporate Symbols

Book of American Trademarks (Vol . 1-9)

How to Improve Your Corporate Identity

by
David E. Carter

Art Direction Book Co.
10 E. 39th Street
New York, N.Y. 10016

Third Printing 1996
Copyright © 1985 by Art Direction Book Co.

All rights reserved. No part of this book may be reproduced or transmitted in any form or by any means, electronic or mechanical, including photocopying, recording or by any information storage and retrieval system, without permission in writing from the Publisher.

Art Direction Book Co.
10 E. 39th Street
New York, NY 10016

ISBN: 0-88108-026-8
LCCC#: 85-8072865

The author wishes to thank the following companies for providing material for use in this book: Deere & Company, The Great Atlantic & Pacific Tea Company, Inc., General Mills, Inc., Pepsico, Inc., Westinghouse Electric Corporation, Holiday Inns, Inc., and Exxon Corporation.

Some of the advertisements and corporate marks in this book are either copyrighted or registered. They are reproduced in this book under the "fair use" provisions of the Federal copyright law, which allows the reprint of such protected material for comment and criticism in books.

*For Christa,
for Lauren,
and most of all,
for Linda.*

Contents

Chapter 1	The Importance of Corporate Identity	7
Chapter 2	Why Most Companies Don't Look as Good as They Are	11
Chapter 3	Improving Your Corporate Identity—Where to Start	17
Chapter 4	Working With a Consultant	22
Chapter 5	The Mark—The Foundation of Corporate Identity	26
Chapter 6	Mark Design—The Thought Process	29
Chapter 7	Testing and Research	36
Chapter 8	Ten Mistakes to Avoid in Mark Design	43
Chapter 9	Typography	56
Chapter 10	Letterheads, Envelopes & Forms	59
Chapter 11	Use of Color	61
Chapter 12	Advertising, Building Exteriors and Rolling Stock	63
Chapter 13	The Importance of Signage	68
Chapter 14	Updating the Corporate Identity	74
Chapter 15	Naming the New Business	86
Chapter 16	Corporate Name Changes	92
Chapter 17	Corporate Identity for Financial Institutions	108
Chapter 18	The Corporate Identity Manual	114
Chapter 19	Costs—Design & Implementation	122
Chapter 20	Putting it All Together Achieving a Consistent Identity	124

Chapter One

The Importance of Corporate Identity
Why Your Company Should *Look* as Good as It Really is

MY PUBLISHER TOLD ME I NEEDED A VERY CATCHY OPENING for this book.

So, I'm sitting in an airport trying to think of a great opening chapter.

I'm in the main terminal, watching the people I see walk by, — and I *immediately* make decisions about those people. I'm amazed how quickly I put people into categories — simply based on their *first visual impression.*

On the next page are pictures of three men waiting for the plane. One I assume is a wealthy businessman. Another looks like a very tired salesman, simply hanging on until retirement, and the third is obviously in the airport by mistake, having escaped from the twilight zone via a time warp.

1 **2** **3**

If a businessman wants to project a highly favorable impression, he'd better look more like Number 1, and less like Numbers 2 and 3.

Now — let's see how this relates to the topic of this book. Just as you and I have made value judgments on three men (all clones, by the way) based on how they *look*, so do people make judgments on *businesses* based on how the businesses look.

The *corporate identity* of a company is its logo (or corporate mark), and the way it is applied to letterheads, signs, trucks, etc., and all visual elements within a company or organization.

Corporate Identity is to a company what clothes are to an executive.

If the corporate identity makes the wrong first impression, the company will lose sales every day — simply because would-be customers don't perceive the company as being what it really is — or as *good* as it really is.

8

For example, take these three company logos (all for the same company). They all present *totally different* corporate images.

Which company looks like it has better management, better products, better service, and indeed, a better future?

Isn't it amazing how quickly you form a first impression — based only on a visual? If your company's first impression doesn't measure up to your firm's realities, you need to make some changes.

This book tells you how to make those changes — *how to make your company look as good as it really is.*

**If you want to make money —
you have to
look like money.**

— Diamond Jim Brady

Chapter Two

Why Most Companies Don't Look as Good as They Are

WHEN IT COMES TO CORPORATE IDENTITY, MOST COMPANIES don't look as good as they really are.

Take a look at the corporate identities of the big Fortune 500 companies. Look at the graphic identities of IBM, RCA, Xerox, Exxon, and others. What you see is a *planned system of corporate graphics, designed to properly portray the company as it wishes to be seen by the public.*

These large, highly successful companies have a corporate identity program that includes (among other things):

1. a well-designed corporate mark
2. a planned consistency to all corporate graphics

The image.

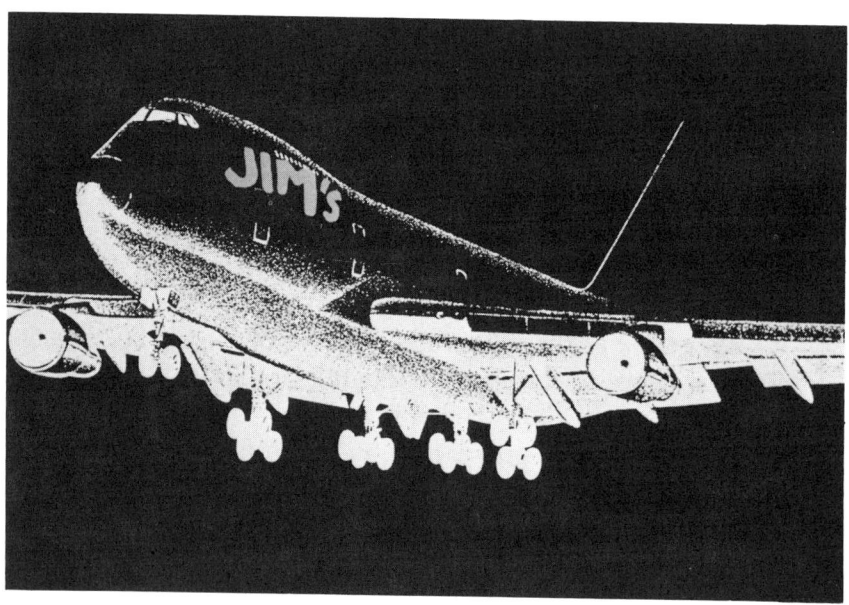

The reality. Based on the image, would you want to buy a ticket on Jim's Airline?

Now, look at most smaller companies. More often than not, you'll find:

1. a poor corporate mark
2. a complete lack of consistency in corporate graphics

When you compare the graphics of the big corporate giants with those of most smaller companies, you see the difference. *To the big company, corporate identity is a very important part of the overall marketing program.*

To most smaller companies, corporate identity is something that "just happened." The logo may have been "designed" in an employee contest, or done by a local "artist" who knew nothing about how detailed corporate identity really is. The stationery was done by the printer (who probably got the job based on having the low bid). And corporate signage was done by the "sign man" who more than likely was also the low bidder.

None of these outsiders really knew (or cared) much about the importance of corporate identity to the company's overall marketing program.

Corporate identity is just as important to the small company — perhaps even more so — than it is to the big company. (I'll tell you more about this later.)

How to tell if your company needs to improve its corporate identity

By this time, you may have already concluded that your company is succeeding in spite of your corporate identity, and that to accomplish your future goals, you'd better make some sweeping changes in your visual images. If that's the case, you may now proceed directly to Chapter 3, for I've already made you a much wiser person, and you are ready to reap untold benefits from the money you've spent on this book.

If, however, you're still a little uncertain as to whether your company looks as good as it really is, I'm going to tell you about two simple tests you can do to find the answer.

The first step is the Logo Test.

Get your company's logo (or corporate mark) and place it with the marks of some really top national companies with good graphics. (If you don't know good graphics, the next page will show you some, just for a starting point. I've even left a blank space where you can put your logo for comparison.) If your mark looks painfully out of place (like some old running shoes next to Gucci loafers) you're a candidate for help. That's test number one.

By the way, if the above test revealed that you're using two or more different logos, you're a victim of corporate schizophrenia, and are a candidate for immediate help — getting *one* corporate mark.

Test two — Gather a number of samples of letterheads, other stationery, forms, photos of signs, trucks, ads, etc. In fact, gather samples of virtually everything which includes your corporate mark. Spread these pieces on a large table and ask yourself this one simple question: "Is everything consistent?"

If the answer is "no," then you're like most companies — your company doesn't actually *look* as good as it really is.

Note: Unless your company has a planned corporate identity program, it's very likely you'll find a complete lack of consistency among the items in this test. I once did a project for a steel manufacturer who had a total of 23 *different* type faces used on various items. Such a lack of consistency costs the company money — and is counter-productive to the overall marketing program.

Does your company need help?

After you've done these two tests, you can make a decision as to whether your company needs help in improving its corporate identity program. If the answer is yes, you may now proceed to Chapter 3 with the firm knowledge that you have made a good decision in buying this book, and you're about to begin a new era for your overall marketing program, because you'll see how much

The logo test

Place your mark here.

These marks are all outstanding. How does your design compare with this group?

corporate identity affects the way your company is perceived by everyone. And — you're going to learn how to make your company look as good as it really is.

If you agree that your company needs help, you may now go to Chapter 3.

(If, on the other hand, you decide that your firm's corporate identity is equal to that of the top companies in the nation, you have no further use for this book, and I hope that you kept your receipt so you can try to get a refund from the bookstore. Or perhaps you can make a charitable contribution to a fellow businessman who is with a company not quite as fortunate as yours.)

Chapter Three

Improving Your Corporate Identity — Where to Start

WELCOME TO CHAPTER THREE. BY BEING HERE, YOU'VE taken a major step toward improving your corporate identity.

If you don't learn anything else from this book, there are two points that should be stressed more than any others:

1. Your corporate identity *must* present a *consistent* look.
2. Your corporate identity *must* present a *quality* look.

It's simple. All you need is a quality look — applied consistently.

On the following page, you'll see examples showing the consistency that is so important to a good corporate identity program. On the opposite page, you'll see the inconsistencies which make up the typical unplanned ID program.

The examples on this page show the consistency which is necessary for a good corporate identity program.

This company has a totally inconsistent graphic look — typical of most companies.

June 16, 1985

Mr. J.B. Bigley
Bigley Corporation
4727 Southern Hills Drive
Chicago, IL 60601

Dear Sir:

Thank you for your recent inquiry about our products. I can assure you that World Wide Wickets produces wickets of the highest quality. You will find wickets of our manufacture in most major corporations.

I am enclosing a brochure that will further explain the versatility of our wickets.

Please call me after you have examined the brochure so we may discuss any questions you may have.

Sincerely,

J. Pierrepont Finch

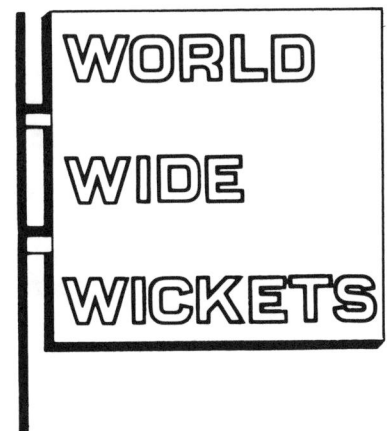

19

Chapter 2 gave you a starting point for comparing your company's corporate mark with that of other companies. That chapter also let you evaluate whether your mark was consistently used.

Now that you've decided you *do* need to improve your corporate identity, the next step is:

Determine the goals of your corporation. Decide what you would like for your company to be in five years. Produce a written statement that includes things such as:

a. size of company (sales $)
b. size of company (number of employees)
c. diversity of company (product range)
d. geographic areas covered
e. position in the minds of the marketplace

An example of this 5-year projection might be:

> a. We would like to have sales of $26 million in 5 years.
> b. We anticipate having a total of 125 employees.
> c. We plan to expand the market for our product (wickets) by producing them in additional sizes and colors.
> d. We will expand from one plant to two (with the new plant to be located on the west coast); we expect to have national distribution through a network of manufacturer's reps.
> e. We expect to overtake our primary competitor and be the nation's number one producer of wickets.

While this 5-year plan is somewhat simplified, it does serve as a beginning point for improving your corporate identity.

Note that this 5-year plan calls for sales growth, product expansion and geographic expansion.

(If your company is planning no growth, no expansion, and indeed, intends to stagnate, return this book, liquidate your company and save yourself the agony of trying to compete in an ever-aggressive marketplace.)

But if your company plans to grow, to progress and to *change* in the next 5 years, an improved corporate identity can be one of the major tools in making this happen.

The above exercise shows you how to verbalize a very simple 5-year growth plan.

Once you have decided what you want your company to become in five years the next step is to create a corporate identity that *looks like the company has already achieved those goals.*

Please go back and read that last paragraph. It may well be the most important words in this book.

Think about what you've just read. I advised you to:

1. Decide what you wanted your company to *become* in five years.
2. Make your company *look like* it's already there.

Deceptive? Certainly not.

If you want to reach those goals, you'd better *look* like you're capable of making it. If you don't have a *corporate look* that equals your goals, you're going to have a much tougher time reaching those goals.

But — if you look like you're already there, it'll make it all the easier to actually achieve the goals.

If you don't believe all this, let's look at an individual who aspires to become a company vice-president in 5 years. In his initial job interview, if he *looks* like a prospective vice-president, he has a much better chance of getting the job.

And, if his everyday appearance is vice-presidential, there's a much better chance he'll reach his goal.

The moral — decide what you want to become, then *look* like you've already made it.

Chapter Four

Working with a Consultant

I DON'T WANT TO MISLEAD YOU, SO LET'S GET RIGHT TO THE POINT. While it's possible for you to improve your corporate identity *in-house*, it's not likely you can do it.

There are a handful of highly qualified corporate identity firms in the nation (see appendix on page 126) and any of these firms can help you greatly.

A good consulting firm will have worked for companies similar to yours in size and in corporate goals. This depth of expertise will be invaluable to you — not only in planning corporate identity, but in helping with other areas of your corporate planning as well.

Corporate Identity in-house: some advice

If you're determined to use this book as a guide on how to improve your corporate identity *in-house*, here are some words of advice.

1. It *has* been done. A few companies with highly qualified design and communications people have produced high-quality ID programs in-house. (Best example: CBS.) But unless you have a very unusual team of employees, it will be unlikely.

2. If you want to have an employee create the new identity program, I highly suggest that you invest in all the books you can find on the topic. If the in-house employee is going to do the project, give him or her all the help you can. (See the bibliography on page 128.)

3. If "saving money" is the prime reason for having an employee do the project, think again. It's my opinion that the really good corporate identity consulting firm will probably *save you more than you pay them* in the long run.

4. Keep in mind that there are only a handful of really qualified corporate identity consulting firms in the USA. It's a complicated field. A "do-it-yourself" CI job is simpler than "do-it-yourself brain surgery," but more complicated than having your summer accounting intern doing your corporate tax return.

By now, you may be saying, "Aha, the man who wrote this book is himself a consultant in corporate identity, and this is but a commercial for his firm's services."
True.
But I also have an obligation to tell you that it's not easy to produce a high-quality CI program. There are literally thousands of do-it-yourself identity programs out there. Most of them look like it. Most of them are abject failures.
If you have ambitious plans for your company, at least talk with one or two good corporate identity firms. Ask them how they work. Ask them about fees. Ask them for samples of their work.

Then, whatever decision you make, at least you've done your homework.

If you have a growing, aggressive company, you are much more likely to reach your goals with a program planned by a top professional in corporate identity.

If you really want to have someone in-house produce the new identity, proceed to Chapter 5 (and please, buy the employee all those books listed in the bibiography.)

If, however, you realize that (despite the fact that I'm a consultant) I have given you some good, brotherly advice — and you've realized that a consultant is the only answer for you, read on.

Finding the right consultant

The appendix on page 126 lists about a dozen top consulting firms. I can recommend any of them without reservations. Each is highly qualified.

(This is not to say that the list is complete. There are certainly other firms in the nation capable of doing high quality work. The list is intended to give the businessman a starting point in his search for a consulting firm.)

How NOT to choose the consultant

1. Call your nephew who had the art class in college.
2. Look in the yellow pages under "artists, commercial."
3. Give the project to an employee who "has a flair for art."

Any of these courses is likely to lead to disaster.

In choosing a consultant, look for a firm that has had experience in working with your size business — one that has a track record in planning and implementing corporate identity projects.

When you're considering a consulting firm, here are some questions to ask them:

> 1. Show me 3 projects which you consider your best work.
> 2. Show me one or two completed projects which are for companies similar in size (and budget) to my company.
> 3. Tell me in 100 words or less your philosophy on how corporate identity affects a firm's overall marketing program.
> 4. What is the background of the people who will be working on my project (should your firm get the job)?
> 5. What is your fee range? How are prices set? Is the price firm?
> 6. How long has your firm been in the corporate identity consulting business?
> 7. Describe the steps you will take — from beginning to end — in planning and implementing a new identity program.
> 8. What is your track record in meeting project deadlines?
> 9. Describe your in-house facilities. How much work (if any) will be sub-contracted?

These questions will give you a starting point, and can be used as a way to compare consulting firms.

I feel these factors are important in choosing a consulting firm:

1. Experience.
2. Personality match.
3. A portfolio of work which really excites you.
4. A fee that you feel is fair.

(There will be more on fees in Chapter 19.)

Using these guidelines, the ideal corporate identity consulting firm is one with a great deal of experience, people who get along with your people, a great portfolio, and the right price.

If you're looking for a consulting firm, that combination will usually result in your satisfaction once the project is over.

Chapter Five

The Mark — The Foundation of Corporate Identity

IN THIS BOOK, A GREAT DEAL OF ATTENTION WILL BE GIVEN TO THE corporate mark.

The reason is simple — the mark is the foundation on which all other aspects of the identity program depend.

Any given *ad* for your company can fail, and the answer is simple — don't run that ad again.

The corporate colors may be removed from any identity program (as they are frequently, when black & white newspaper ads are run) and the program does not have negative effects.

The size relationship between the mark and the signature (name in type) can be changed, and little happens.

But take away the corporate design — the mark — and the corporate identity disappears.

The mark is the single unifying element in the corporate identity program.

And because the mark is included in all phases of the corporate identity, *it must be good.*

If your corporate mark doesn't properly represent your company, it's time to make a change.

If you're ready to make that change, the next chapter will tell you all about starting the design process.

**A good corporate mark
tells the world —
"this company is a *success*...
we're going places."**

Chapter Six

Mark Design — The Thought Process

THIS CHAPTER IS ABOUT THE PROCESS OF DESIGNING THE trademark, or mark, or logo or signature.

These terms are often used interchangably, to mean a corporate identity element. In this book, you'll find that I have used these terms somewhat loosely.

Definitions

Signature: The corporate name set in type; normally used consistently in a particular typographic style.

Mark: A *design* used consistently as a visual symbol to represent a company or organization. It may or may not be used with a signature.

Mark used without signature

Mark used with signature

Wordmark: A mark which also incorporates the corporate name *into the design.* It is impossible to physically separate the two elements and still have a design remaining. (The signature and mark *can* be separated, and often are.)

Since we're still in the *thought* process, let's look at what makes a good corporate mark.

What makes a good corporate mark.

1. Original and distinctive (There are more than 10,000 *registered* marks using a diamond shape.)
2. Legible
3. Simple
4. Memorable
5. Easily associated with the company
6. Easily adaptable for all graphic media

If you use that 6-line check list to compare any design you'll ever encounter, chances are you'll be able to select designs that are effective.

Now, let's get back to the thinking process that must take place before ideas can be put onto paper.

First of all, it's important to know as much as possible about the client. (See the section in Chapter 3 about determining the goals of the corporation.)

Once you are satisfied that you know as much as possible about the company, its goals, etc., you can begin *thinking* about designing a corporate mark. (The designing will come later. For now, let's *think*.)

I find that it helps to put these items on paper.

With corporate goals very much in mind, I write down items such as:

1. Defining the uses of the mark. (Will it ever be animated for TV, for example. If the answer is yes, that's important in the design process.) Will the mark ever be used on trucks, signage and other large formats? If so, this will affect the design approach.

2. How important are budget considerations? Will printing budgets allow for a multi-colored mark on stationery, for example? (A taxpayer-supported institution can rarely afford a 2-or-3-color letterhead, for example.)

3. What image is the company trying to project: conservative, flambouyant, old-line, high-tech, etc.? The answer further redefines the direction of design thought.

You'll probably discover other parameters of your own to add to the above list. (Each project is different, so different needs will appear.)

The word list

Once the above questions have all been answered (or considered), the next step is to make a *word list*. The word list is simply a list of words that somehow associate with the client company. The list is used in the design stage as a springboard for ideas.

To show you what a word list might look like, let's take a set of facts for a client company, then compile a word list for it.

The company is Robinson Excavating, a large company with more than 100 machines on various jobs, moving dirt at construction sites.

Our word list consisted of:

> bulldozers
> movement
> powerful
> earth moving
> big trucks
> digging ditches
> transport
> dump trucks
> landscape
> sculpt
> maneuver

Once the word list was completed, we began the design process. Here you see some of the rough sketches which were involved in getting to a final solution.

33

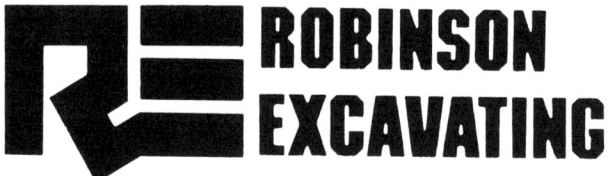

The final solution — the one now in use by the client — is this. We feel it is very appropriate for the client's type of business. The colors — brown and green — reflect the earth and grass that are moved by the machines.

We have not shown all the rough designs that led to the final step, but you can see enough to give you some idea as to the evolutionary process that created the final mark.

***Consistent* use
of your corporate mark
is the key to immediate recognition
for your company.**

Chapter Seven

Testing and Research

ONCE SOME PROGRESS HAS BEEN MADE ON GETTING A NEW name and/or corporate mark, many people will want to do some sort of sophisticated "test" on the new idea.

Here are some of the typical "tests" that are done by novices:

1. Take the new mark (and color scheme) home to the wife. Ask her what she thinks of the colors. (Do they match the new draperies?)

2. Call in the secretary (who had an art class in junior high school) and ask for her opinion. Flaunting her B+ in Psychology 101, she unwinds with a 17-minute monolog on the hidden meanings of the phallic symbols the designer has so obviously "hidden" in the logo.

3. Call in the board of directors, (all 12 of them) and ask them to vote on which of the designs will be chosen for final use. A half-hour tape of this exercise could win an Emmy for comedy. I've seen some of the funniest comments (thoroughly unintentional) when this method of "testing" is used. This exercise is also guaranteed to bring out the very worst, most childish, traits among your directors.

4. Call in the secretarial pool, plus some of the executives. Ask them "Which logo do you prefer?" Count the seconds until someone responds with "Which one do *you* like, Mr. Bigley?"

As you have probably guessed by now, none of the four examples are really "testing."

What is testing?

Testing is a procedure that attempts—using some variation of the scientific method—to determine which item would most likely meet the standards of the target group.

For example, if you're testing a proposed new logo that's going to have to appeal to consumers, you should NOT ask:

1. company officers
2. company spouses
3. secretaries
4. directors
5. and other similar groups

If you want to know what you consumers think, try something revolutionary: *ask your consumers.*

There are several easy ways to measure consumer attitudes toward a new mark (or name).

Focus groups are composed of a cross-section of people *who are representative of the groups whose opinions you value.* For example, if your project wants to know how plumbers feel about your new corporate identity, your group should be composed of a random sample of plumbers in your market area.

In a focus group, a moderator leads the discussion and the topic of conversation is directed toward those items which are being studied. (If you're sincerely interested in learning more about this type of testing, go to your public library where you'll find some good books on the topic.)

Another type test is to present your new material one-on-one with a random sampling of the group whose opinion you find valuable. Various tests can be done by mail, in the office, in home, etc. These tests can measure responses on a 1-5 scale, on a positive/negative scale, and many others.

The sample chart below can be adapted for many uses in corporate identity testing.

It can be used to test individuals' reactions to proposed logos, names, colors, typography, and other aspects of the corporate identity program.

An example of such a test is shown on the following page in which two different logos have been tested.

	Very	Somewhat	Neither	Somewhat	Very	
Exciting	☐	☐	☐	☐	☐	Dull
Old	☐	☐	☐	☐	☐	Young
Active	☐	☐	☐	☐	☐	Passive
Masculine	☐	☐	☐	☐	☐	Feminine
Refined	☐	☐	☐	☐	☐	Crude
Expensive	☐	☐	☐	☐	☐	Cheap
Stylish	☐	☐	☐	☐	☐	Dowdy
Friendly	☐	☐	☐	☐	☐	Unfriendly
Beautiful	☐	☐	☐	☐	☐	Ugly
Industrial	☐	☐	☐	☐	☐	Fashionable
High Class	☐	☐	☐	☐	☐	Low Class

At the conclusion of the test, the numbers are tabulated for each response, and lines are drawn to connect the most frequent response for each line.

These tests can serve as good guidelines in decision making in corporate programs. I emphasize that they are guidelines, not commands. Use the test results along with a good dose of common sense for best results.

SMITH

	Very	Somewhat	Neither	Somewhat	Very	
Exciting	12	61	17	8	2	Dull
Old	0	12	51	12	3	Young
Active	20	71	9	10	0	Passive
Masculine	5	55	24	9	7	Feminine
Refined	42	18	35	4	1	Crude
Expensive	5	69	15	7	4	Cheap
Stylish	16	59	20	1	3	Dowdy
Friendly	65	20	9	6	0	Unfriendly
Beautiful	14	14	48	11	13	Ugly
Industrial	9	65	19	7	0	Fashionable
High Class	58	22	14	5	1	Low Class

SMITH

	Very	Somewhat	Neither	Somewhat	Very	
Exciting	4	1	10	58	27	Dull
Old	11	13	25	36	15	Young
Active	5	10	18	19	48	Passive
Masculine	14	19	13	34	20	Feminine
Refined	6	11	32	29	22	Crude
Expensive	2	15	21	19	43	Cheap
Stylish	1	6	9	18	66	Dowdy
Friendly	3	0	32	41	24	Unfriendly
Beautiful	0	5	34	17	44	Ugly
Industrial	8	13	69	7	3	Fashionable
High Class	12	15	14	26	33	Low Class

Why some firms insist on testing

In many companies, testing is a way of life. All new products must undergo extensive testing. All new ads must be tested. And, certainly, any new identity must be tested.

Why so much reliance on testing?

It's simple. When a great deal of money is at stake, *high-paying jobs* are at stake. It makes sense, therefore, that the MBAs must have someone else to blame, in case things go wrong.

In the event of a disaster, they can always say "...but the tests said we should do it. Don't blame *me*."

If you think I'm exaggerating on my opinion of a major reason for testing, ask anyone in a big company where MBAs run the show. You'll discover that my little needle is actually being very kind.

So—do we test or not?

Let's assume you have a potential new corporate identity program on paper. You like it. The ad people like it. The legal people have checked it out—they like it because they see no conflict. Yes—even your wife liked it. (You didn't ask, but she told you anyway.) Actually, she "kinda" liked it, which may be taken as an overwhelming vote of confidence.

You have the new design applied to mock-up signs. It appears in color on dummy letterheads, calling cards, packages, etc. A large photo of the new design is shown on one of the firm's delivery trucks.

Everyone likes it. (It *can* happen.)

Do you test, or believe everyone's gut feeling?

That's *your* final decision to make, and no $18 book is going to presume to take that decision-making power away from you.

But, since I'm writing about the topic of *to test or not to test*, let me give you the benefit of my experience.

In 6 specific instances regarding corporate name changes, the client was so concerned* about making the right decision that he insisted that a test be done *by an outside firm*, "to eliminate any bias we might have."

I'm happy to report that in all 6 cases, our first choice was annointed by the outside groups as THE right choice.

Our "gut feeling" was confirmed by the time-consuming, dollar-eating studies. What we "felt" was confirmed by "conclusive data" on the part of the researchers.

I genuinely believe that the "sixth sense" of a highly-qualified consulting firm can be more accurate than any test yet devised by man or machine. I believe that tests *can* serve as guidelines. But *don't* let a test dictate your final decision.

*footnote—see scared

Testing—the other side of the story

I don't want to come across as being anti-testing (we've used it in many cases) but I do want to let everyone know that testing is not the panacea for the communication's world's questions.

One oft-told story about the pitfalls of testing. A large company insisted that all the TV spots to air on the network be tested in advance of appearance. One spot failed to make the minimum score, but the client decided to run the spot anyway. (His gut feeling was "I like it; I'll run it.")

Even though the spot—according to tests—should never have run on TV, it went on to become a Clio-winning commercial.

The best commercial of the year was a failure by testing—yet a smash hit in reality.

The moral—if you *do* decide to test your corporate identity program, do so with this fact in mind—the instinct of a professional can be just as valuable as cold numbers on paper...maybe more so.

A final word about testing:

The Coca-Cola Company spent huge amounts of money to test the decision to eliminate the "old" formula for Coke and replace it with a new, sweeter, formula.

All the tests said "do it."

"New" Coke was introduced. The nearly 100 year-old Coca-Cola formula was changed.

Within weeks, consumer demand for "old Coke" forced the company to revive the old brand as "Coca-Cola Classic."

Don't assume that testing can tell you everything you want to know.

Chapter Eight

Ten Mistakes to Avoid in Mark Design

EVERY YEAR, THOUSANDS OF NEW MARKS ARE DESIGNED. MOST of these suffer from some major flaw—a mistake so bad that it virtually assures that the mark won't be fully effective in its goal: making the company look as good as it really is. There are ten primary mistakes that cause a mark to fail.

Each of the ten mistakes will be discussed briefly here, and you'll see examples to illustrate why they fail.

1. Lines too thin
2. Depends on color to be successful
3. Use of near-abstract initial for first letter of name
4. Not appropriate for the type of business
5. Wrong proportions for most uses
6. Too busy
7. Fad type face
8. Uses visual cliches
9. Complete lack of imagination (Incredibly Dull)
10. Grade-school design (the 6th grade solution)

Reading over this list and looking at these examples isn't going to make you into a top mark designer. It *will* give you some very important guidelines to follow in the process of improving your corporate identity.

On the following pages are examples of the ten mistakes to avoid. On each page, three bad examples are shown on the left; on the right a better design solution is shown.

Lines too thin

*The logos on the left illustrate design problems.
On the right are design solutions which work well.*

Thin lines do not reproduce well. Avoid them.

Depends on color to be successful

*The logos on the left illustrate design problems.
On the right are design solutions which work well.*

*The designs on the left depend heavily upon color.
If the design MUST have color to be good - it isn't good.*

Use of near-abstract initial for first letter of name

The logos on the left illustrate design problems.
On the right are design solutions which work well.

 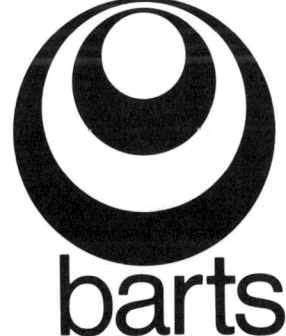

This tricky visual only confuses the reader.

Not appropriate for the type of business

The logos on the left illustrate design problems.
On the right are design solutions which work well.

Don't try to get cute - be appropriate.

Wrong proportions for most uses

The logos on the left illustrate design problems.
On the right are design solutions which work well.

Marks which are extremely vertical or extremely horizontal will NOT work in most applications.

Too busy—too many elements

The logos on the left illustrate design problems.
On the right are design solutions which work well.

Need I say more?

Grade-school design

*The logos on the left illustrate design problems.
On the right are design solutions which work well.*

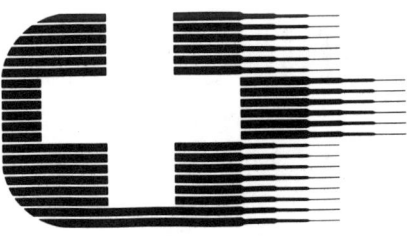

Have the people who did these designs on the left ever see the inside of a junior high school?

Fad type face

The logos on the left illustrate design problems.
On the right are design solutions which work well.

A fad typeface such as these on the left will produce a design that is quickly out of date.

Uses visual cliches

*The logos on the left illustrate design problems.
On the right are design solutions which work well.*

Avoid the obvious solution.

Lack of imagination

The logos on the left illustrate design problems.
On the right are design solutions which work well.

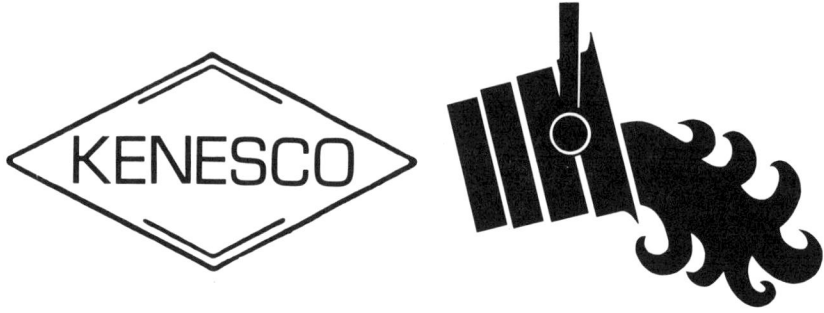

These "designs" on the left show a lack of imagination.

Your corporate mark speaks for you.
It may say,
"strong, growing and professional."
Or, it may say,
"small time, with a dubious future."

The *right* corporate mark
can change your firm's
first impression greatly.

Chapter Nine

Typography

TYPOGRAPHY IS A VERY IMPORTANT PART OF ANY CORPORATE identity program, for the type style(s) used will communicate a great deal about the company, just as the accessories worn by an individual communicate a lot about that person.

There are literally thousands of type face variations, and some of the major faces can produce moods, just as colors can create moods and indicate characteristics.

For example, take the adjectives on the following page, which might describe companies or their products:

creative

old-line

high-technology

heavy industry

elegant

expensive

durable

With the right typeface, each of these words takes on an added dimension as shown below.

CREATIVE

old-line

high-technology

HEAVY INDUSTRY

elegant

expensive

durable

To summarize this short chapter, when typefaces are chosen, work to get a design that is appropriate for the company and its products. (Translation: Don't do things like this:)

Steel Manufacturer

Jewelry Store

COMPUTER

Chapter Ten

Letterheads, Envelopes & Forms

JUST ABOUT EVERY COMPANY USES LETTERHEADS, ENVELOPES AND forms.

So for most companies, much of the "first impression" made by the business will be made by stationery, invoices, business cards, etc.

Assuming that the company already has a good corporate mark, here are the simple steps to follow to get the maximum benefit from printed matter:

1. Select a single color and texture of paper for use in letterheads, envelopes, cards, etc. This will not only give you the consistent look that is so important, it will also save money (by allowing you to order in quantity).

2. Use the corporate mark consistently on all printed items. That is, place the mark in a consistent position, with similar size relationships between mark and type, etc.

3. If color is an important part of the corporate identity system, use it on all important pieces, such as letterheads, envelopes, business cards, etc. Use of color is less important on items such as packing lists, internal forms, etc.

4. For important executive-level communication, the use of monarch-size stationery is especially recommended. This stationery may use embossing and/or engraving, in order to create the quality image desired.

5. The layout of the letterhead and envelope must be planned so that a good typing format is possible.

This envelope does not give the impression of a modern, high-tech engineering firm.

Chapter Eleven

Use of Color

COLOR CAN BE A VERY EFFECTIVE TOOL IN THE CORPORATE identity program.

In fact, very few planned identity programs *do not* have color as part of the system. (You seldom see a corporate identity program totally in black and white.)

Since color will likely be a part of your program, let's see some of the considerations:

1. Colors create moods. Red means passion, power; blue indicates coolness; yellow indicates weakness. There are entire books devoted to the psychological meanings of colors. The effects of colors upon emotion should be considered in planning corporate colors.

2. For many products, the use of color is almost dictated by the product itself (i.e. orange juice, steel, greenhouse, etc.).

3. When color is part of the identity system, it will not always be used. For example, most newspaper ads will have a black & white mark. Many internal forms, shipping notices, invoices, etc., may appear in black ink. Be certain that the mark works as well in black & white as it does in color.

4. Multiple colors cost. Most corporate identity systems use black and *one* additional color. The primary reason (besides simplicity) is cost. Printing two or three colors in addition to black can add a substantial amount to a printing bill. Add this up over the years, and it comes to a substantial amount — just to get a second color (besides black) in your corporate identity program. Best advice is — go with black and one color. It will probably be more effective — and you'll save money.

5. It is most important to use your corporate color consistently. There is a universal color matching system to help you avoid problems in matching colors. The most widely used system is the Pantone Matching System. Nearly any printer can receive your request for PMS 286 and provide you with the exact color you want. (There are literally hundreds of different shades within the Pantone Matching System.) Make a PMS color part of the identity system, and you'll be assured of having a consistent match on every job you print.

Chapter Twelve

Advertising, Building Exteriors and Rolling Stock

ONCE YOU HAVE A WORKABLE CORPORATE MARK, IT IS important that it appears *consistently* on virtually every item — from business cards to huge outdoor signs on buildings.

Advertising

Since your advertising will (hopefully) be seen by prospects to buy your product, it is very important that your corporate mark be highly visible in all your ads. It is also important that the mark be in the same position in all ads.

Research has shown that the ideal place for the corporate mark in print ads is in the *lower right corner*. The next best place is for the mark to be *centered in the lower one-quarter* of the ads.

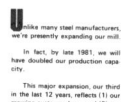

1 **2**

Research shows the best place for the mark to appear is (1) the lower right corner, and (2) centered in the lower quarter of the ad.

Building exteriors

If your building, factory or other locations are on heavily traveled highways, you have an added advantage of using your exterior space for corporate identity.

If you are lucky enough to be on a heavily traveled road, make the most of this situation.

Use the corporate name and mark to identify your location. Over a year's time, it'll be seen literally several million times. The identity can be actually painted on the side of the building, or you can have signs constructed to fit the site. Either way, you'll greatly enhance your corporate image in your home region with this type of identity.

If you have heavy traffic near your building, take advantage of this by identifying your company to the millions of people who will pass by over a period of time.

Rolling stock

This refers to all company automobiles, trucks, trains, planes, etc.

It is important to have all rolling stock identified with the corporate mark, since to do otherwise is to miss the opportunity to display your corporate mark and name *many millions of times a year.*

Let's assume that you have only one semi-trailer truck on the road for 10 hours a day, 260 days a year. That totals 2,600 hours a year, or 156,000 minutes.

Now, if that truck (with your corporate mark on it) is seen by 20 people per minute (a slow day on the interstate) *your logo on that one truck will be seen 3,120,000 times in a year.*

That's just what one truck can do for you. If you have a fleet, you're talking about making tens of millions of impressions with your logo each year.

If you have company cars, trucks, trains or planes, your corporate mark will give you millions of free impressions every year.

Be Consistent

We recently completed an identity project for a small business with a fleet of 14 trucks. Each truck had the company name on it, but each was a different color truck, and there was no consistency to the type style used in painting the name on the trucks.

When the changeover was completed, the response was amazing. The company president told us that he had numerous comments such as "...you must have expanded. I see your trucks all over the place." "I didn't know you were so large. Your trucks are everywhere."

The number of trucks didn't change. The only change was to make all the trucks look alike. The visual impression made by the trucks was instant and very positive. People *remembered* seeing the trucks, because they all looked alike. (Before, all the trucks looked different, and no one remembered seeing them.)

Chapter Thirteen

The Importance of Signage

FOR MANY TYPES OF BUSINESSES, OUTDOOR SIGNS ARE THE KEY TO attracting customers. Firms such as motels, restaurants, etc., must have effective signs in order to attract the first-time customer.

Let's look at some examples of how signage can work. Both signs are for an "off-brand" hotel. But one repels, while the other is inviting.

These motel signs give you an idea of what to expect inside.

These signs for two pizza restaurants show a marked contrast.

Take a good look at both signs, pretend you're driving through a strange town, have an insatiable craving for pizza, and are suddenly confronted with two pizza parlors. The signs quickly become more than just identification — they represent the restaurants' *quality of product.*

On a 1-5 scale (1 being greasy, 5 being succulent) the sign on the left probably gets a 1. In addition, the sign on the left looks like a place you *don't* want to take your family.

While the differences in *quality of image* are readily apparent, note that the *basic elements* on the two signs are the same. The primary change is in the design. Both say "we sell pizza."

Quite simply, the two signs, like the people in Chapter 1, convey starkly different images, with relatively small changes.

Signs for financial institutions

When a family moves to a new town, it's time to choose a new bank. When other factors are fairly equal, the image of the bank can be a deciding factor on where to open a new account.

Look at these two examples of outdoor signs of two competing banks. Where are you most likely to open an account? Which bank looks more progressive, more financially secure?

Staid look

Dynamic look

If you're involved in banking, take a good look at your signs — and those of your competition. You can't afford signs that don't reflect well upon your bank. (For more on corporate identity for financial institutions, see Chapter 17.)

Visibility of signs

While general design principles should be followed when it comes to signs, one additional point is very important — signs MUST be visible from long distances. This includes using typography that is readable from a long distance. Compare these two examples to see how important the right type is to signage.

When people are driving (especially at night) it is easy to drive past a turn-off before the sign becomes readable.

Further, if two signs are side-by-side, and one can be seen and read from twice as far away as the other, that sign will make *twice* the impact over any period of time as the inferior sign. Translate that into dollars and look at the results:

Assuming 6 million impressions a year (not unusual for a sign):

Sign A

Readable from x distance
gets 6 million impressions
total visibility time: 36 million seconds

Sign B

Readable from 2x distance
gets 6 million impressions
total visibility time: 72 million seconds

With no additional cost of sign production, Sign B achieves 18 million seconds more visibility time. It costs nothing more — yet it produces twice the readership as Sign A.

Those 36 million free seconds of exposure time — which cost nothing — would nearly equal the cost of a commercial inside the *Tonight Show Starring Johnny Carson*. In effect, this one well-designed sign is getting you extra exposure worth well over $10,000 a year — simply because you had the good fortune to have read this book and chose your sign (or sign designer) with care.

McDonald's—best signage in the USA

Perhaps the best example of a sign being well-designed to meet this need for visibility is McDonald's. Notice how the McDonald's sign here stands out.

Even manufacturers need good signage

While the need for good signage is obvious for retail businesses, even manufacturers and other locations that don't make direct sales also should have a good signage program.

Why? Because that location will want to attract high-quality employees, it will want to *present the image of a community asset*, and even though no sales are made there, the company should nevertheless present its best possible image...everywhere.

In summary — if signs are a part of your business, get the maximum exposure by having signs that are readable from a long distance, and ones that reflect favorably on the company.

**For many businesses —
hotels, restaurants, etc.,
signs are perhaps the most important
advertising medium.**

**The traveler looking
for quality food or lodging
seeks a quality image
in signage.**

Chapter Fourteen

Updating the Corporate Identity

STYLES AND FASHIONS CHANGE. WHILE A GOOD MARK IS designed to last a number of years, it sometimes becomes necessary to make a change.

While the specific reasons for making a change may vary, the only real reason for a change in identity is to bring the company's image more in line with its present reality. Another common reason for a change in corporate identity is that the mark was poorly planned in the beginning.

Making the decision to update the corporate identity may be simple — or it may be undertaken as a study project to determine the advisability of making such a change. Either way, when a change is decided upon, one basic question is paramount — *how much change is needed?*

Equity

When deliberating the question of "how much change?" it becomes important to consider the *equity* of the existing mark. In corporate identity, equity refers to the positive (or negative) image produced by a mark by its exposure over the years. Just as you build equity in a house as you pay off the mortgage, so do you build equity in a mark as you expose it to the public over the years. However, just as it's possible for a deteriorating house to have *negative equity*, a mark can have the same problem. In other words, the publicity value of a mark can be a negative factor in producing sales. For example, a mark for an engineering company which shows a slide rule indicates a very backward company. In this case, the mark has *negative equity*, and no attempt should be made to retain any part of the design.

When a business has a fairly well-accepted mark that has reflected favorably on the company, that mark probably has some positive equity. In such cases, the new design should retain some visual tie in with the old mark.

For example, in 1975 the great Atlantic & Pacific Tea Company, Inc., wanted to make some visual changes which would better indicate changes which were taking place within the A&P organization, so the company adopted its new logo.

While the new mark was designed, in the words of the company, "to express our forward progress," the new design reflected the past, with a certain amount of the old mark being evident in the new one.

OLD NEW

1876

1912

1936

1937

1950

1956

1968

For some old, well-established firms, the corporate mark has been re-evaluated and re-designed (with equity retained) on a periodic basis.

Deere & Company, a major manufacturer of farm implements, had its original mark designed in 1876. Today the current version is the 7th generation of the original. Each successive version shows the link to the past.

Similarly, the Westinghouse Electric Corporation has a mark that is the 6th in a line of descendents from the original which was designed in 1900.

1900 - 1910

1910 - 1922

1922 - 1940

1940 - 1953

1953 - 1960

1960 - Present

In 1983, Holiday Inns Inc., began a change which would eventually eliminate its long-familiar "great sign" and replace it with a simpler, less costly identification.

A Holiday Inn press release on the change explained the major reasons for the change — "We are changing our sign in order to project a more contemporary image that better reflects our hotel chain's current range of property types, customer base and product quality."

In the decade after the first Holiday Inn hotel was built in 1952, the company's primary customers were families traveling by automobile. The Great Sign, so named because of its imposing size and high roadside visibility, served to generate business from passers-by, who originally represented nearly 95% of the chain's business.

Holiday Inn hotel's popularity with business travelers, however, has increased significantly since the early 1970s. Today, more than 85% of Holiday Inn hotel roomnights are sold to business travelers or to couples traveling without children, and less than 3% of customers are "walk-ins" without reservations.

Since the mid-1970s, the Holiday Inn hotel system has responded to this shift in customer base and travel patterns, as well as to the increasing expectations of today's travelers, by upgrading the quality of its product through new development, refurbishments, and the removal of hotels that do not meet the chain's strict standards and market requirements.

The Great Sign, which was designed for America's first interstate highways as a means of promoting the company's early two-story, U-shaped roadside motels, now is found in a variety of location types ranging from high-rise urban properties to resort locations in 56 countries and territories throughout the world.

Holiday Inn is one of the world's most famous trademarks, and while we recognized the need to modernize our image, it was important that we retain our Holiday Inn logo.

The solution was to retain its packaging.

The new sign, which was a T-shaped frame topped by a round edged rectangular sign face, was created by S & O Consultants in San Francisco. A national sample of lodgers who had spent four or more nights in a hotel during the previous 12 months indicated they preferred the new sign on the basis of "overall appeal" by a three-to-one margin over Holiday Inns' existing sign.

ORIGINAL "GREAT SIGN"—The Holiday Inn sign was designed in the 1950s for the burgeoning Holiday Inn chain as a means of attracting passers-by in the nation's then-new interstate highway system. The sign has 426 incandescent bulbs, 836 feet of neon tubing, and sixteen 96 foot fluorescent tubes in the attraction panel. It was dubbed the "Great Sign" because of its imposing size and high roadside visibility.

NEW HOLIDAY INN SIGN—Holiday Inns' replacement for its original sign features a modern, rectangular design intended to project a more contemporary image and to better reflect the international hotel chain's current quality of product, customer base, and range of property types. The new version also will be more cost and energy efficient.

Aesthetics aside, the new sign incorporates considerable practical improvements. Gone are the 836 feet of neon tubing and 426 incandescent bulbs; in their place is more subdued fluorescent backlighting that is projected to be 65-75% more energy-efficient. According to company officials, average annual maintenance costs should be reduced by at least 55%, and the cost of erecting the new sign is estimated to be 34% less than that required for the Great Sign — $23,000 compared with the current $25,000.

Assuming an annual energy inflation rate of 15% and maintenance increases of 10% annually, the new sign will pay for itself in energy and maintenance savings in approximately five years.

In the case of Holiday Inn, the change was dictated by a drastic change in the firm's customer base — over the years, a much greater percentage of the firm's customers had advance reservations. There was no need for such a large (and costly) identifying sign.

In each of the previous examples, the new design contained some link with the past — making use of the positive equity of the old mark.

However, when the old mark is simply a bad design, the usual solution is to visually sever all ties with the past and come up with a totally new look.

When Kentucky Electric Steel was founded in 1964, the company was literally started on a shoestring. The primary concern was to get the plant into production. Things such as visual identity were of no importance to management at that time.

However, after the company had been in business for about two years, management felt that a logo was needed. The budget was still tight, so a compromise was reached. A contest was held for employees (and their children) to design a company logo.

Management announced the contest in its mimeographed employee newsletter, and promised a $25 savings bond for the employee (or child) who came up with the winning design.

(This approach to designing a logo doesn't usually produce especially good marks, but it does offer one consolation — the savings bond costs only $18.75 or so.)

The winning entry was chosen, the designer (?) was pleased to receive the savings bond, and the company had a logo at last. (See illustration)

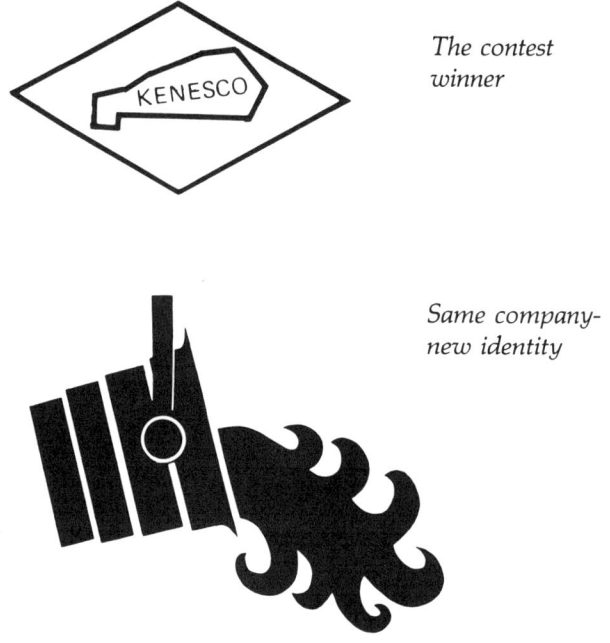

The contest winner

Same company— new identity

In this case, the company was trying to compete with firms such as US Steel, Armco and others. The company was quickly becoming a qualified competitor, but the *image* it projected didn't do justice to the steel company's product.

After the logo had been in use for two years, we were retained to come up with a new design. In this case, the question of equity never came up. The existing mark was so bad, we felt it was mandatory to make a drastic break with the visual past.

While the mark we did was a little more costly than the original savings bond design, it helped project the image of a growing, high-technology steel manufacturer.

Two last examples of maintaining equity before we move on to the next chapter.

The photos here show how General Mills has changed the image of Betty Crocker over the years.

1936

1955

1965

1968

1972

1980

And finally, here's how Pepsi-Cola has updated its corporate identity to keep pace with the times.

*Approximately
1898-1901*

*Approximately
1905-1906*

*Registered 1937-
probably used
back into the
late 20's*

1950-script

*Approximately
1962-1969*

present

Summary

If a company stays in business for an extended period of time, at some point an update of the identity will probably be necessary.

When that times comes, the new design will either be (a) a total departure from the past, or (b) a design that retains the equity of the previous mark. Sound marketing thought must go into this decision.

Chapter Fifteen

Naming the New Business

FEW PEOPLE GET THE CHANCE TO NAME A NEW BUSINESS. WHEN the opportunity comes, beware, for many traps await you. The primary trap to avoid is the *I thought too small* trap. Thinking small means you never really considered that the firm would:

1. grow beyond a limited geographic range.

2. grow beyond its original product range.

Here are a few examples of companies which were named and immediately were candidates for the "I thought too small trap:"

Greenup County Bank (didn't consider crossing a county line with a branch).

Michigan Waterbeds (didn't consider becoming a full-line furniture store in several states).

Diecasters, Inc. (didn't consider that the firm would ever do anything other than diecastings).

The bad thing about the *I thought too small trap* is that you may be in the trap for years—with no reason to escape. Since you have no reason to escape, you don't even realize you're in the trap *until escaping is an immediate necessity.*

For example, the 20th Street Bank was a great name for a bank—once upon a time. Back in the days when this particular state had very strict banking rules that prohibited *any* form of branch banking, it was a descriptive name that helped people immediately know the bank's location.

For many years, the 20th Street Bank was in the trap—but didn't know it, because it never had a reason to try to escape.

Along comes new banking laws, allowing not only *branch* banks, but laws which permitted statewide banking.

Can you imagine a 20th Street Bank trying to have a 12th Street Branch, or a Clarke Street branch? Worse still, picture the 20th Street Bank opening an office in a town with no 20th Street. Or, worse again, 20th Street Bank opening a branch in a town *with* a 20th Street, but with the bank located—naturally—elsewhere.

As you can see, the *I thought too small trap* can have some serious results—and they usually come many years after the business was founded, and often come as a result of growth.

Once the business realizes that its name is a negative, it's time to escape that trap. If you're in a hurry to know how to escape, turn to Chapter 16 now.

There *is* a solution

If you're reading about all this with sweat beads on your forehead and a nervous twitch because *your* company has a name that is no longer suitable, we have a solution: turn to Chapter 16—Corporate Name Changes.

One last word about new corporate names

If you file papers to incorporate, you'll be dealing with the Secretary of State in the state in which you file for your corporate charter.

In filing for your corporate name, you will find that your first choice for a name may have already been taken by someone else. You also may find that your second, third, fourth and fifth choices for a name have *also* been taken.

Be prepared for the possibility that your chosen name—the "perfect" name for your corporation—will be turned down by the powers that be.

Once you have your name in place legally, and your company starts to grow, you face one more potential problem.

Should you physically open an office in another state, you may have to register with the office of the Secretary of State.

Let's assume that you've been doing business from your New Mexico base as *Gizmo, Inc.* for five years, and expand into Colorado, you may be required to *register* in Colorado.

If you suddenly discover that there's a *Gizmo Corporation* in Colorado, your application may be denied.

(Nobody said this was going to be easy.)

At this point, I should point out that I'm not an attorney, and I'm not trying to practice law without a license, and the best way to avoid such traps is to get the help of a good attorney when you deal with such things as incorporating a business, and expanding a business to another state. Our wonderful 50-state system has a few pitfalls, and this is one that the growing business has to avoid.

Avoid the "alphabet soup" syndrome

When naming a new business, it's a simple matter to avoid all the problems mentioned above by using initials.

Let's assume that Charlie and Dave get together and decide to call their new business C&D Enterprises. After all, it doesn't limit the business to a geographic location or to a product line. And—Charlie and Dave get to have the ego trip of using their initials in the company name.

Before you decide to use initials for your company's name—DON'T.

If you do use initials for the company name, you're falling into a large vat of "alphabet soup" and you will be smothered by much larger companies who have the money to force their names upon the public consciousness simply by virtue of tons of mass communications.

Consider that *all* the firms listed below are taken from a recent *Fortune 500* list, and you'll see how many *big* companies use initials for corporate names.

GE	CDC	NL	CF
GM	CPC	SCM	VF
IBM	IC	IMC	NVF
USS	NCR	CBI	ACF
ITT	FMC	AMF	GAF
PPG	P&G	GATX	AM
RJR	BFG	USI	M/A
LTV	AMAX	BASF	MEI
TRW			ITW

Now, consider that all these firms are *huge*. They are all national companies, with big marketing budgets. All, in theory, have the clout to make their name known to virtually every American adult.

Yet, once you get past the halfway mark on the list, you'll probably begin having trouble knowing who the company is.

With that in mind, take a minute to think how much impact you'll have to make as C&D Enterprises, or whatever, in order to have any serious dent in the minds of the public.

In summary, when you're naming a new business,
1. Think big
2. Plan ahead

Would a company called "J & D Book Co." have grown as rapidly as one called "Century Publishing?"

**Probably not.
Pick a name that *sounds* established.**

Chapter Sixteen

Corporate Name Changes

WITHIN THE LAST 25 YEARS, THERE HAVE BEEN ENORMOUS changes in the structure of American business. Before 1960, "conglomerate" was an unknown word in the world of industry. Few businesses ventured outside their main area of expertise. Today, it's not uncommon for a company to own a number of widely diverse divisions, making old corporate names totally unacceptable for marketing products to the masses.

In addition to the changes mentioned above, many smaller businesses are finding that their geographic horizons have changed *The 20th Street Bank* may have been a grand old name back in 1920, but how will it stand now that deregulation means that banks can have branches statewide in many cases?

There are any number of reasons why a company might need

to undergo a name change, and 10 such reasons are outlined in this chapter. But just so you won't look at this chapter head and say "My company will never change *its* name . . . why Grandfather would turn over in his grave," let's take a quick look at some companies — big and small — which have undergone corporate name changes.

Old Name	*New Name*
Dayton Malleable Iron Co.	Amcast Industries
Cities Service	Citgo
Esso	Exxon
International Pipe and Ceramics	Interpace
Clinton County Farm Bureau Co-op	AgMax
Farmer's Union Central Exchange	Cenex
Associated Seed Growers	Asgrow
First National Bank of New York	Citibank
Alabama Bancorporation	AmSouth
National Bank of Commerce of Seattle	Ranier Bank
United Aircraft	United Technologies
Allied Chemical Corporation	Allied Corporation
Corn Products Corporation	CPC
R J Reynolds Tobacco	RJR
Haloid — Xerox	Xerox
Rexall Drug & Chemcial Co.	Dart Industries Corp.
American Brake & Shoe Foundry	Abex Corporation
Radio Corporation of America	RCA
Pittsburgh Plate Glass Co.	PPG
Armco Steel Corporation	Armco Inc.
Tennessee Gas Transmission Co.	Tenneco, Inc.
Ashland Oil & Refining Co.	Ashland Oil, Inc.
South Penn Oil Co.	Pennzoil Co.
Mary Carter Paint Co.	Resorts International
Citizens Savings	Civic Savings
A B Dick Products of W. Va.	Quorum Corporation
Begley Drug Company	Begley Company

This is just a small fraction of the companies which have undergone corporate name changes in recent years. In fact, nearly 1,000 different corporate name changes have been recorded in the USA each year in the 1980s.

As you can see from the list above, some of the changes are small alterations in wording; others are somewhat drastic. In every case, however, the corporate management felt the need was great to have a corporate name change.

In each case, the management decided that the long-term needs of the company would be better served by having a different name.

Below are ten reasons for a corporate name change. If your company fits into any of these categories, it's time to consider a name change.

Ten reasons for a Corporate name change

1. Name is too long
2. Name is geographically restrictive
3. Name is outdated in terminology
4. Name is too limiting in range of operations
5. Name is misleading
6. Name is difficult to pronounce
7. Name is difficult to remember
8. Name is associated with failures
9. Name is not appropriate for multinational use
10. Name change dictated by merger

You can look at the list of companies which have changed their name (earlier in this chapter) and see examples of many of these reasons.

In order to give you a little more insight on name changes, below is a brief discussion of the "reason-why" for several name change projects we've managed.

ANDERSON FEDERAL
SAVINGS & LOAN ASSOCIATION
Before

American Federal
Savings & Loan Association
After

1. *Anderson Federal Savings & Loan Association*, Anderson, Indiana. The firm has been a fixture in the Anderson market for years, and its name was a plus—as long as it confined its business to Anderson. But the firm's aggressive management wanted to expand greatly — and realized that the name "Anderson Federal" would be a liability in other towns. The result was a name change — to American Federal.

First National Bank of Greater Miami

Before

After

2. *First National Bank of Greater Miami*, Miami, Florida, was expanding beyond the greater Miami market . . . even into New York City. It needed a name that would position it against the nation's major banks without sounding like a regional bank. The result — a new name: Consolidated Bank.

Before

Family
Care
Outpatient
Center

After  John Marshall
Medical Services

3. *Family Care Outpatient Center,* a teaching arm of the Marshall University Medical School was offering high-quality medical care from physicians on the medical school teaching staff. However, the "outpatient center" in the name drove away affluent prospective patients who perceived the quality of care as low. In addition, the name did not reflect the close ties to the medical school. The new name — John Marshall Medical Services — attracted the upper-income patients and solved the communications problems.

Before

After

4. *Union Bank,* Tucson, Arizona. New ownership of the bank was not able to change the image of the prior group of owners, even though three years had passed since the acquisition. A new name — Interwest Bank of Arizona — immediately communicated that it was a new beginning for the bank.

Before

Mid★America Federal

After

5. *Ohio Federal, First Federal* and *State Fidelity Savings* of Ohio all merged into one of the nation's largest financial institutions. Since it was a true merger with all three being equal partners, a new name was a necessity. In addition, future considerations were given to the possibility of having offices in adjoining states. The new name — Mid-America Federal — presented the right image for future expansion.

Before

After

6. *Clinton County Farm Bureau Cooperative Association,* of Frankfort, Indiana was much more than a local farmers' co-op. It was a major force in agriculture in the Midwest, but the name positioned it as just another local county farm bureau. The new name — AgMax — solved the problem.

PRODUCTS COMPANY

Before

After

7. *A B Dick Products Corporation of West Virginia* had been an A.B. Dick Products distributor for years. But as time changed, the company began selling Xerox copiers, computers, and even graphics arts supplies. A.B. Dick represented only a small fraction of the firm's sales. The old name was a hinderance in making sales. The new name — Quorum Corporation — solved the problem.

If it seems as though more than half the space on case histories of name changes is devoted to financial institutions, it's because more than half the name changes in the USA each year are done by financial institutions.

One last word about name changes. I have been associated directly with more than 40 corporate name changes. We have found that when properly planned and executed, a corporate name change means *an increase in business.* Don't fear a corporate name change. Make it into an event — a marketing tool for now and the future.

The Exxon name change

In 1972, the largest name change in the history of American business took place when Standard Oil Company (New Jersey) became Exxon Corporation.

The following material was provided by Exxon's Public Affairs department, which gives a great deal of insight about the name change project.

Standard Oil Company (New Jersey) moves to become Exxon Corporation

The Board of Directors of Standard Oil Company (New Jersey) adopted a resolution recommending to its shareholders that the company's name be changed to Exxon Corporation. The company called a special meeting of shareholders to vote on the resolution.

Also announced was the decision of Esso Chemical Company Inc. and Enjay Chemical Company to use Exxon as their primary trademark and name in the U.S. This followed the announcement by Humble Oil & Refining Company, Jersey's principal domestic affiliate, that it would change to Exxon.

At the same time, Jersey announced its intention to realign the company's organization in the U.S., by merging Humble, as well as Esso Chemical and Enjay, into Exxon Corporation, with the approval of the shareholders.

J.K. Jamieson, Jersey board chairman, said at the time that the decision was made that the new name would simplify and improve company communications and identification in the U.S. by linking the corporation with its principal affiliates and their products and services, as well as with its shareholders and employees. "This is important for competitive reasons," said Jamieson. "Unlike some leading companies in our business, our operations in the U.S. have been conducted by corporate subsidiaries using various names and trademarks. This was largely a result of court injunctions against Jersey's use of the name 'Standard Oil' or 'Esso' in 20 states.

"Our own Standard Oil Company (New Jersey) name has served us well and is rich in tradition," said Mr. Jamieson. "But it has also led people to confuse us with some of our competitors. To solve this problem once and for all, we decided to start from scratch with a new name we could use anywhere.

"Our choice, Exxon, was a distinctive name—strong, modern, easily recognized. Humble's tests had shown it to be effective in the marketplace. And the use of Exxon as a corporate name by Jersey as well would eliminate the need for separate corporations for our principal business activities in the U.S., thus making possible a somewhat simpler, more flexible corporate organization."

Mr. Jamieson stressed that the merger would not represent any modification in Jersey's basic policy of decentralized management. "The U.S. oil and gas operations, a very significant part of Jersey's assets, would continue to require the attention of a strong and self-reliant management group," he said. "Likewise a strong, decentralized management group for our chemical interests continued to be desirable. Thus the responsibilities of the managements of these companies would remain essentially unchanged. And there were no plans to change office locations."

Looking to the future, Mr. Jamieson said, they were confident that Exxon was the name with which they could build a really distinctive national identity for the company.

Following are details of the name change and corporate organization plans:

> Standard Oil Company (New Jersey) became Exxon Corporation.
>
> Humble Oil & Refining Company became Exxon Company, U.S.A. (a division of Exxon Corporation). Its subsidiaries with "Humble" in their names, such as Humble Pipe Line Company, changed to Exxon; others' names did not change.
>
> Esso Chemical Company Inc., responsible for Jersey's chemical business, and Enjay Chemical Company, its affiliate, became respectively, Exxon Chemical Company (a division of Exxon Corporation) and Exxon Chemical Company, U.S.A. (a division of Exxon Chemical Company).
>
> The Exxon Chemical trademark became similar to the rectangular red, white and blue design selected for Exxon's domestic petroleum business.
>
> Jersey Enterprises Inc., a Jersey affiliate responsible for new business enterprises, became Exxon Enterprises Inc. Jersey Nuclear Company became Exxon Nuclear Company.

The company's chief executive emphasized that, since there was no impediment to the use of Esso abroad, affiliates outside the U.S. would continue to use it.

Why switch to Exxon?

In 1911, the Supreme Court handed down a ruling resulting in the breakup of the Standard Oil organization as it then existed. This action led to developments that culminated in Jersey's decision to change its name.

Of the 34 competing companies formed from the former Standard Oil organization, seven retained the name Standard Oil in their company titles and used it in certain areas of the country for marketing; Standard Oil of California for major territories in the west; Standard of Indiana for the midwest, except Nebraska, which was retained by Standard of Nebraska; Standard of Ohio for that state; Standard of Kentucky for the southeastern states; Standard of New York for the northeast, and Standard Oil Company (New Jersey) for the Middle Atlantic states.

Jersey introduced its Esso brand name in 1926 and by 1933 was using it in 18 states from Maine to Louisiana. In 1935, the Esso brand name was introduced into the St. Louis area, and Indiana Standard took court action challenging its use there. This was the first of a series of law suits with companies that retained the Standard Oil name which resulted in blocking Jersey's use of Esso outside the eastern area.

In 1937, a Federal Court prohibited the use of the Esso trademark in 15 midwestern states.

Beginning in 1960 the Enco brand name was introduced into the western U.S. and into other states where Esso was not used. In 1961, California Standard acquired Standard of Kentucky, following which Jersey introduced the Esso brand name into the southeast. A declaratory judgment by the Federal District Court in Mississippi upheld Jersey's right to do this, but that judgment was reversed by the Federal Court of Appeals in 1966, and the Supreme Court declined to hear a petition for certiorari. The Federal Court in St. Louis in the same year refused to change the 1937 St. Louis ruling barring the use of Esso in the midwest.

In 1967, Enco was introduced into the five southeastern states covered by the court ban on Esso.

Jersey's legal appeals on these rulings were exhausted in 1969 when the Supreme Court declined to review the decision in the St. Louis case, and it was at that point that the company made the firm decision to look for another name.

The name EXXON was chosen after extensive research and market testing showed it to be a strong one, quickly recognized, easily remembered and highly acceptable to customers.

The change to EXXON would simplify and improve company communications and identification in the U.S. by linking the corporation with its principal affiliates and their products and services, as well as with its shareholders and employees.

How do you change your name?

If you're operating a business and decide to change its corporate name, how do you decide what the new name is going to be?

Most companies making a change retain at least some of the elements in the old name. A few change the name completely. Still fewer not only coin a new name but a new word. This was the case with Standard Oil Company (New Jersey) which recommended to stockholders that Jersey's name be changed to EXXON.

An informal survey of changes by corporations in the United States in recent years disclosed that they may be divided into five general categories. Not necessarily in order of frequency, they are:

1. *Acronyms or syllable combinations*: Among these are Pennzoil Company, which used to be South Penn Oil Company, and Tenneco, Inc., which formerly was Tennessee Gas

Transmission Co.

2. *Name changed to a group of initials*: These are perhaps some of the best known new versions of names. They include such names as PPG (for Pittsburgh Plate Glass Co.) and RCA Corp. (for Radio Corporation of America).

3. *A complete change from the old to new names*: Many companies throw out their old name completely and start over again. Thus, American Brake Shoe Co. became Abex Corp., and Rexall Drug & Chemical Co. became Dart Industries Corp.

4. *A change to eliminate product, service or geographic identification*: In this category, R.J. Reynolds Tobacco Co. became (R.J.) Reynolds Industries, Inc. and Libby-Owens-Ford Glass Co. dropped the word "Glass" to become simply Libby-Owens-Ford Company.

5. *A change that retains at least one word of the old name*: This type of change occurs fairly frequently for a variety of reasons. Examples are Cudahy Packing Company, which shifted to Cudahy Co., and Socony Mobil Oil Co., which changed to Mobil Oil Corp.

The reasons for deciding which way to go in the questions of a name change are almost as varied as the changes themselves. But relatively seldom does a company change because of legal restrictions on its name or trademark.

Exxon highlights

Exxon is an invented word, chosen because it best meets a series of qualifications for a corporate and brand name. It is short, quickly recognized, easily remembered and just as easy to pronounce. The double-X, except for proper names such as Foxx, occurs only in the Maltese language.

Trademark clearance was one of the most detailed aspects of the preparatory work for the name change. Researchers checked trademarks and company names in all 50 states and the District of Columbia to learn if any similar names were already in use in this country. In the process, they even examined over 15,000 telephone directories.

Selection of the EXXON name involved research over a period of more than three years. The world's principal languages were studied to avoid undesirable meanings or connotations, as well as avoiding difficult pronounciations that might bother foreign-born customers.

Thousands of names, some produced by computers, were considered at the start. The number was reduced to 234, then to 16, then to eight, which were studied and tested until all but EXXON were eliminated. Enco, one of the company's current U.S. trademarks, was rejected in part because in Japanese it can mean "stalled car."

When the company was organized almost 90 years ago, it took the name "Standard Oil of New Jersey." About 10 years later, this was changed to "Standard Oil Company," which is still the legal name of the company. After 1911, when a Supreme Court ruling resulted in the breakup of the organization, and seven companies retained the words "Standard Oil" in their names, the name normally has been written "Standard Oil Company (New Jersey)" to distinguish it from the other six.

As a result of the 1911 decree, the Standard Oil organization was divided into 34 separate companies. Seven of them continued to use the Standard Oil name. By 1972, four major companies retained the Standard Oil name. They are: Standard Oil Company (New Jersey), The Standard Oil Company (Ohio), Standard Oil Company (Indiana) and Standard Oil Company of California. In addition, there is Standard Oil Company (Kentucky), which was acquired by Standard of California in 1961 but continues to operate under its own name.

Among the trademarks used at various times in the company's 90-year history are: Esso, Enco, Enjay, Humble, Polarine, Stanacola, Aladdin, Security Oil, Eupion and Astral Oil. Esso, the best known of all, was introduced in 1926.

Approximately 224 million shares of the company are now held by some 780,000 shareholders. There was no change in the corporate name on the outstanding stock certificates for these shares until they were transferred. As new certificates were issued, they bore the EXXON name.

In the U.S., more than 25,000 service stations underwent the name change. The stations were served by the company's chief domestic affiliate, Humble Oil & Refining Company, which became EXXON Company.

When your company name is no longer a company asset— change it.

Don't worry about the consequences. *Well-planned* name changes result in *increased* business.

Chapter Seventeen

Corporate Identity for Financial Institutions
Special Considerations for a Special Market

NOT SO MANY YEARS AGO, WHEN LIFE WAS SIMPLE, BANKS AND savings and loans had a license to make a profit. These institutions offered only passbook savings (usually at 4 or 4½% interest).

If your credit was good, and you applied for a loan, it would cost you 6 or 6½% interest. (When interest rates raised to 7%, people were up in arms and used words like "exorbitant" and "gouging.")

Life in the financial industry is no longer quite so simple. Whereas once banks lived in peaceful co-existence with savings &

loans (some even had overlapping directors) now there is a dogfight going on. De-regulation, combined with changing times, has brought a number of new competitors into the arena. Brokerage houses, credit unions, mortgage companies, small loan companies and others are all competing for the same business today.

And, before this decade is complete, it's likely that major national retail chains will be heavily involved in the financial business.

To make the matter more complex, the financial industry is one of two in the nation which touches virtually every adult. (The other is food, and the list stops there.)

Because financial institutions serve virtually *everyone*, and because there is so much new competition for business, *it is essential that an aggressive financial institution present a visual image that will enable it to capture its desired market share.* Because of the complexity of the financial marketplace, financial institutions have very special corporate identity needs.

This chapter deals with the special problems of financial institutions relating to corporate identity.

Since nearly everyone is a prospect for the services offered by a bank, savings & loan or similar institution, mass marketing techniques are commonly used, even by financial institutions in small markets.

Once we accept the fact that the food business is the only other industry which reaches so broad a prospect group as financial institutions, we should look at what the packaged food industry is doing that might be emulated by banks and S&Ls.

Packaged foods offer a variety of products—just as financial institutions now do with their many services.

Packaged foods offer convenience — just as banks do with their flexible hours, drive-in locations and 24-hour teller machines.

Packaged foods offer a service necessary for physical survival — just as banks offer a service necessary for financial survival.

And the manufacturers of packaged foods spend a great deal of time and money in packaging their product right. Just as consumers make quick decisions on buying packaged food due to a *visual impression*, people often make decisions on trying a financial institution on a visual impression.

Compare the appeal of these packages of dog food.

Generic package

Brand food design

The generic package says "low budget," and instantly raises questions about quality.

Now compare the instant images produced by these two logos for a savings & loan association. One looks very staid, lifeless and unimaginative. The other shows progress, vitality and financial stability. We can even imagine how different the officers and directors of these two institutions might be.

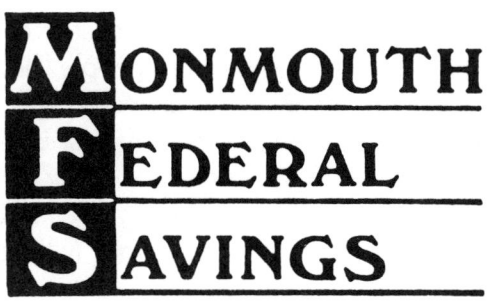

The American consumer is on the move, with 20% of the population changing addresses each year. This fact alone means that a great deal of any bank's customer base is somewhat volatile and subject to change.

Add to this the number of changes of account location made by customers due to dissatisfaction, desire for better or additional services, etc., and you see that as much as 25% to 30% of a bank's customer list is subject to turnover each year.

When a person chooses a new bank, (or S&L or similar firm) the decision is often based on *image*. The customer perceives the newly-selected bank as having friendlier people, being more progressive, giving quicker service, or whatever.

And much of that image is formed by first impression — by the bank's advertising, and its corporate identity.

Pretend for a minute that you've just moved into a new town and are driving near the shopping center that will be nearest your new home.

You're going to establish a relationship with a financial institution, but you haven't really been exposed to much advertising for any banks or S&Ls.

As you're driving, you see two banks on opposite corners of the street. Each location is equally convenient. Each building looks similar. The only difference is in the signage, which is shown here.

Which bank looks more progressive? Which looks like a place that's going to be able to give you a wide range of service?

While both banks may *in reality* have similar service, etc., the BankWest sign gives the impression of a much more progressive, dynamic bank than the First Peoples Bank signage.

For a financial institution — perhaps more than any local business, *looking* right is a primary step to attracting a desired customer base.

P.S. After this book had been typeset I was driving down Lankershim Blvd. in Los Angeles and saw a sign that made me stop the car.

"Here is the future of banking," I thought. Here's the reason for *every* bank in the nation to re-evaluate its own corporate identity and corporate image.

The sign: SEARS SAVINGS BANK. The nation's best-known retailer is in the banking business, and as regulations are loosened, there exists the possibility that every Sears store in the country could be a branch bank. Just think of it. Millions of people already have a Sears credit card. Their credit has already been established. And Sears obviously has strong financial resources.

In the future, your bank is going to be competing with Sears Savings Bank, and perhaps the J.C. Penney Bank, the K-Mart Bank, and a lot of others.

The message: Plan your corporate identity NOW so you will be in a position to compete favorably against this type competitor.

Chapter Eighteen

The Corporate Identity Manual

ONCE THE COMPLETE PACKAGE OF CORPORATE GRAPHICS IS established, all the information is usually committed to paper.

For many companies, that information becomes a *Corporate Identity Manual*, which is a detailed "dress code" for corporate graphics.

The corporate identity manual shows various applications of the corporate mark, use of color, typography, and other elements which make up the complete identity program.

In most cases, the corporate identity manual is done in a ring binder. This allows changes and additions to be made when appropriate.

The front of the corporate identity manual should include a letter from the firm's chief executive officer. This helps to convey the importance the company places in the corporate identity program.

Copy in an identity manual should be held to a minimum. Since the topic is visual, examples should be visual, with captions used to describe important points.

Copies of the manual are normally given to purchasing people, suppliers, and anyone concerned with specifying or producing items bearing company identification.

Excerpts from a corporate identity manual for AgMax, an Indiana-based agricultural cooperative follow. (The original manual was 8½ x 11, with most pages in color.)

Introduction 2

Elements of identification

Corporation
The **legal name** of the company is AgMax, Inc. Wherever the name is legally required to be on forms, packages, documents, etc., the full legal name should be used.

The **communicative name** of the company is **AgMax**. This is the name which should be used on communications that are not legal, such as stationery, signs, advertising, etc.

The **corporate mark** of the company is the design shown above. The design has a number of guidelines concerning color, size, etc. which are to insure the proper usage of the mark. These guidelines are included in this Corporate Identity Manual.

Signage 20

Signage 21

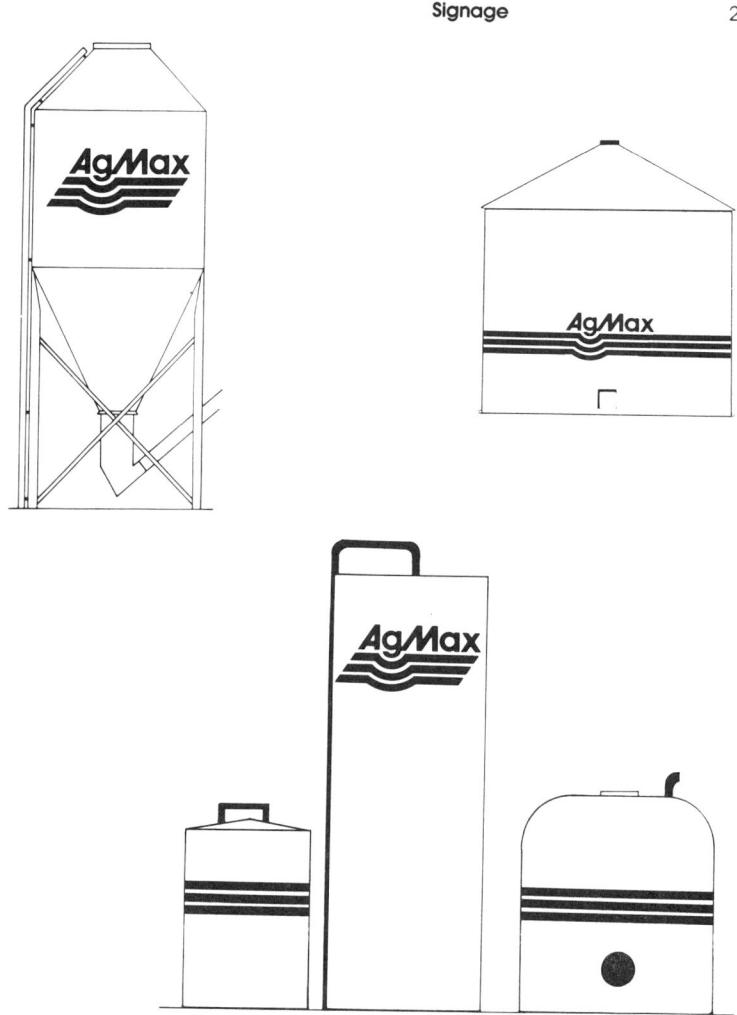

Typography/Signatures

ABCDEFGHIJKLMNOPQRSTUVWXYZ&
abcdefghijklmnopqrstuvwxyz
1234567890(.,:;!?"—/$-%)A ACA Ⓒ
EA EA Fr GA HT LA MN TCE Rr RA SS ST TH UT VV Wev wy

ABCDEFGHIJKLMNOPQRSTUVWXYZ&
abcdefghijklmnopqrstuvwxyz
1234567890(.,:;!?"-/$-%)

ABCDEFGHIJKLMNOPQRSTUVWXYZ&
abcdefghijklmnopqrstuvwxyz
1234567890(.,:;!?"-/$-%)

ABCDEFGHIJKLMNOPQRSTUVWXYZ&
abcdefghijklmnopqrstuvwxyz
1234567890(.,:;!?"-/$-%)

Typography

Since the **AgMax** mark is based on typography, it is important that all type be coordinated whenever possible.

The type faces shown on this page are approved. The four different weights of Avant Garde Gothic allow for a great degree of latitude in design, while maintaining the desired corporate look.

Signature

Whenever the name **AgMax** is set in type in the context of an advertisement, brochure or other commercial communications piece, it is desirable to have the name set as follows: **AgMax**. This is done by using Avant Garde Gothic Bold, with special characters for the **A** and **M**, with appropriate letterspacing to allow the **g** and **M** to overlap as shown here **gM**.

Text copy in this instance is Avant Garde Gothic Book, the lightest weight approved.

Advertising 15

Advertising

All **AgMax** advertising is to carry the corporate mark in one of two locations: (1) centered at the bottom of the ad OR (2) in the lower right one-quarter of the ad.

This is the logo position which most enhances the effectiveness of an ad's name recognition.

In newspaper advertisements, the mark will normally be run in black and white. For color magazine advertisements, the mark should be run in the standard Pantone® colors.

Typography

Advertising typography should be one of the four weights of Avant Garde Gothic: book, medium, demi, or bold.

Vehicles 31

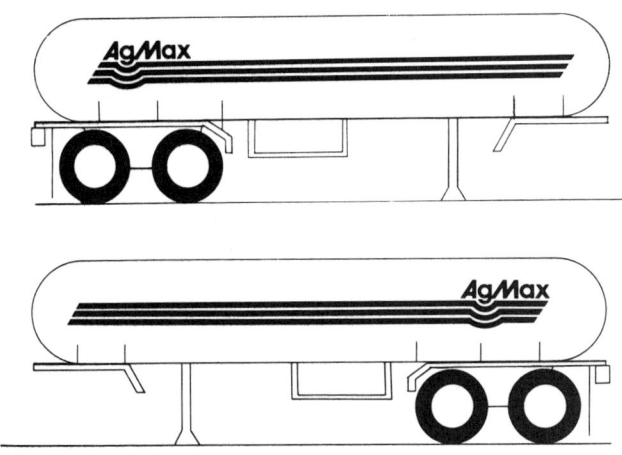
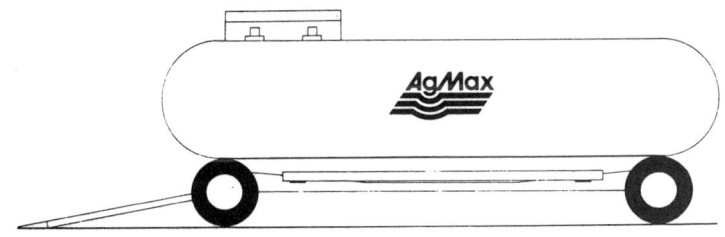

Chapter Nineteen

Costs — Design & Implementation

MANY COMPANIES HAVE RETAINED OUTDATED CORPORATE identities due to a fear about implementation costs.

These people assume that since corporate identity includes *all* printed forms, *all* signage, *all* rolling stock, — virtually *everything* with the company name on it — that the costs of implementing the program will be astronomical.

Indeed, if everything with the company name were discarded overnight, the costs would be huge.

However, most companies introduce the new identity in a transition period. This phasing-in of the new identity program has two purposes:

1. Costs are reduced greatly. The new identity is applied to items as they must be replenished. (The next time invoices are ordered, the new design is used, etc.)

2. There is some evidence that communications is enhanced when the new identity is phased in over a period of months. That is, there is a stronger connection between the old and new identities when they overlap.

Since most items involving corporate identity must be replaced periodically, there is no extra cost involved in implementing the new identity to these items.

The only items that would need to be replaced for identification purposes only would be signage, which normally has a life of 5-8 years.

Consultant fees

Since implementation of the identity program can be done with relatively little additional cost, the primary non-recurring cost is that of the consulting firm.

The range of consulting fees can be considerable. Depending upon the size of the project, the fee can range from several thousand dollars to a high of somewhere into six figures.

But by the time a firm is ready for a new identity, the cost of *doing* the program is a minor consideration.

The primary concern is the cost to the company's future of *not presenting a corporate image that matches the company's abilities.*

The good consultant will lead the client company through all the problems that can arise in the design and implementation of a new identity program.

And, quite frequently, the consultant will save the client company an amount of money greater than the consultant's fee, simply because of the consultant's expertise in knowing sources and knowing the pitfalls that make the costs escalate well beyond the amount budgeted.

Chapter Twenty

Putting it All Together. Achieving a Consistent Identity

BY NOW, YOU HAVE A GOOD CORPORATE MARK — ONE THAT adequately reflects your company, its abilities and its goals.

You may even have a corporate manual, in which you have a set of graphic standards to assure that everything is presented in a planned manner.

Now — the most important thing you can do is to be sure that *your corporate identity becomes a consistent program.*

You must use only one corporate mark (don't be tempted to resurrect that old mark you buried recently, for any reason).

You must use only the approved corporate colors. (Never specify medium blue, when what you really want is PMS 286.)

You must use one particular type face for the corporate signature. (Never settle for something "almost the same" just because the printer didn't have your typeface in stock.)

In short, you must do everything to see that your company presents a consistent visual image wherever the company name appears.

If you do have a corporate identity manual, *use it*. Appoint an *Internal Coordinator* in your company to monitor all use of the corporate mark to see that everything is consistent.

If you don't have a corporate identity manual, at least get a few pages outlining the general principles on the use of corporate graphics.

Be certain that all people involved in the purchasing of items with the company name on it are aware of the importance of corporate identity.

A short, two-hour session of the importance of corporate identity would be a good way to show a group in the company that you are committed to achieving a consistent visual identity.

Once you have a good graphics system, you can have a consistent identity by:

1. Letting everyone know that top management is behind it.
2. Appointing someone to make certain that consistency is achieved. And if an inconsistent use appears, a short note will probably prevent a future occurrence.

And, finally, if anyone questions the need for a consistent corporate identity, offer to lend them this book.

Consulting Firms

Primo Angeli Graphics, 285-c Connecticut St., San Francisco, CA 94107 (415)974-6100

Anspach Grossman Portugal Inc., 711 Third Ave., New York, NY 10017

Babcock & Schmid Associates, 3689 Ira Rd., Bath OH 44210 (216)666-8826

David E. Carter, Inc., 1505 Carter Ave., Ashland, KY 41101 (606)329-0077

DiCristo & Slagle Design, 741 North Milwaukee St., Milwaukee, WI 53202 (414)273-0980

Follis Design, 2124 Venice Blvd., Los Angeles, CA 90006 (213)735-1283

Fulton & Partners, Inc., 330 West 42nd St., New York, NY 10036 (212)695-1625

Goldsholl: Design & Film Companies, 420 Frontage Rd., Northfield, IL 60093 (312)446-8300

The Graphics Studio/Gerry Rosentswieg, 811 N. Highland Ave., Los Angeles, CA 90038 (213)466-2666

Malcolm Grear Designers, Inc., 391 Eddy St., Providence, RI 02903 (401)331-2891

Nash Hernandez Graphic Design, 1140 Empire Central, Suite 200, Dallas, TX 75247 (214)630-2575

Landor Associates, Pier 5, Ferryboat Klamath, San Francisco, CA 94111 (415)955-1200

Lee & Young Communications Inc., One Park Ave., New York, NY 10016 (212)689-4000

Congdon Macdonald Inc., 420 Lexington Ave., New York, NY 10170 (212)697-9300

Lippincott & Margulies, Inc., 499 Park Ave., New York, NY 10022 (212)832-3000

Overlock Howe Consulting Group, Inc., 4484 West Pine Blvd., St. Louis, MO 63108 (314)533-4484

Roth Co., 535 N. Michigan Ave., 2312, Chicago, IL 60611 (312)467-0140

Robert Miles Runyan & Associates, 200 E. Culver Blvd., Playa del Rey, CA 90293 (213)823-0975

Pat Taylor Inc., 3540 S St. NW, Washington, DC 20007 (202)338-0962

The Weller Institute for the Cure of Design, 2240 Monarch Drive, Park City, UT 84060 (801)649-9859; 2427 Park Oak Dr., Los Angeles, CA 90068 (213)467-4576

Other Books of Interest

Barach, Arnold B.; Famous American Trademarks, Public Affairs Press, Washington, DC, 1971

Capitman, Barbara Baer; American Trademark Designs, Dover Publications, Inc., New York, 1976

Carter, David E.; Corporate Identity Manuals, Art Direction Book Company, New York, 1976

Carter, David E.; Letterheads (Vols. 1-5), Art Direction Book Company, New York, Annual

Carter, David E.; Designing Corporate Identity Programs for Small Corporations, Art Direction Book Company, New York, 1982

Carter, David E.; Evolution of Design, Art Direction Book Company, New York, 1983

Carter, David E.; Logo International, Art Direction Book Company, New York, 1984

Carter, David E.; Designing Corporate Symbols, Art Direction Book Company, New York, 1975

Carter, David E.; Book of American Trade Marks,(Vols. 1-9) Art Direction Book Company, New York, Annual

Cocomas Committee; Corporate Design & Systems, Typony, Inc., New York, 1979

Cocomas Committee; Basic Design Elements and Their Systems (Vols. 1-9), Sanno Institute of Business Administration, Tokyo, 1979

Cooper, Al; World of Logotypes (Annual), Art Direction Book Company, New York, 1982

Crosby, Fletcher & Forbes; A Sign Systems Manual, Praeger Publishers, Inc., New York, 1982

Diethelm, Walter; Signet Signal Symbol, Editions ABC, Zurich, 1972

Eksoll, Olle; Corporate Design Programs, Rein Publishing Corporation, New York, 1967

Follis, John & Dave Mammer; Architectural Signing and Graphics, Whityen Library of Design, New York, 1979

Henrion, F.H.K. & Alan Parkin; Design Coordination and Corporate Image, Reinhold Publishing Corporation, New York, 1967

Herdeg, Walter; Trademarks and Symbols, Graphis Press, Zurich, 1948

Jacobson, Egbert; Trademark Design, Paul Theobald, Chicago, 1952

Kamekura, Yosako; Trademarks and Symbols of the World, Litton Educational Publishing Company, 1965

Kamekura, Yasaku; Trademark Designs of the World, Dover Publications, Inc., New York, 1981

Kuwayama, Yasaburo; Trade Marks & Symbols,
Vol. 1: Alphabetical Designs, Van Nostrand Reinhold Company, New York, 1973

Kuwayama, Yasaburo; Trade Marks & Symbols, Vol. 2: Symbolical Designs, Van Nostrand Reinhold Company, New York, 1973

Margulies, Walter P.; Packaging Power, World Publishing Company, 1970

Marquis, Harold H.; The Changing Corporate Image, American Management Association, New York, 1970

Mulcahy, Bob; The Trademarks (TM), Society of Typographic Arts, Chicago, 1968

Rand, Paul; The Trademarks of Paul Rand, George Witten Born, Inc., New York, 1960

Ricci, Franco Maria and Corinna Ferrari; Top Symbols & Trademarks of the World (Vols. 1-8), Deco Press, Milan, Italy, 1973

Ries, Al and Jack Trout; Positioning: The Battle For Your Mind, McGraw-Hill, New York, 1981

Room, Adrian; Dictionary of Trade Name Origins, Routledge & Kegan Paul LTD., Boston, 1982

Rosen, Ben; The Corporate Search For Visual Identity, Van Nostrand Reinhold Company, New York, 1970

Sandgren, Russell A. and Joseph M. Murtha; Bank Identification Planning, Bank Marketing Association, Chicago, 1974

Vivarelli, Carlo; The Principles in the Design of Trademarks, Neue Graphik, Switzerland, 1960

Wildbur, Peter; International Trademark Design, Van Nostrand Reinhold Company, New York, 1979

Wildbur, Peter; Trademarks: A Handbook of International Designs, Reinhold Publishing Corporation, New York, 1966